AF266695

K-2 Math
Volume 1

© 2013 Todd Deluca
OnBoard Academics, Inc
Newburyport, MA 01950
800-596-3175
www.onboardacademics.com

ISBN: 978-1-939796-72-1

ALL RIGHTS RESERVED. This book contains material protected under International and Federal Copyright Laws and Treaties. Any unauthorized reprint or use of this material is prohibited. No part of this book may be reproduced or transmitted in any form or by any means, electronic or mechanical, including photocopying, recording, or by any information storage and retrieval system without express written permission from the author / publisher. The author grants teacher the right to print copies for their students. This is limited to students that the teacher teachers directly. This permission to print is strictly limited and under no circumstances can copies may be made for use by other teachers, parents or persons who are not students of the book's owner.

Table of Contents

Skip Counting

Key Vocabulary

Skip Counting

Multiply

Multiple

Skip Counting by Twos.
Write the number of boxes in the space provided and then count by twos.

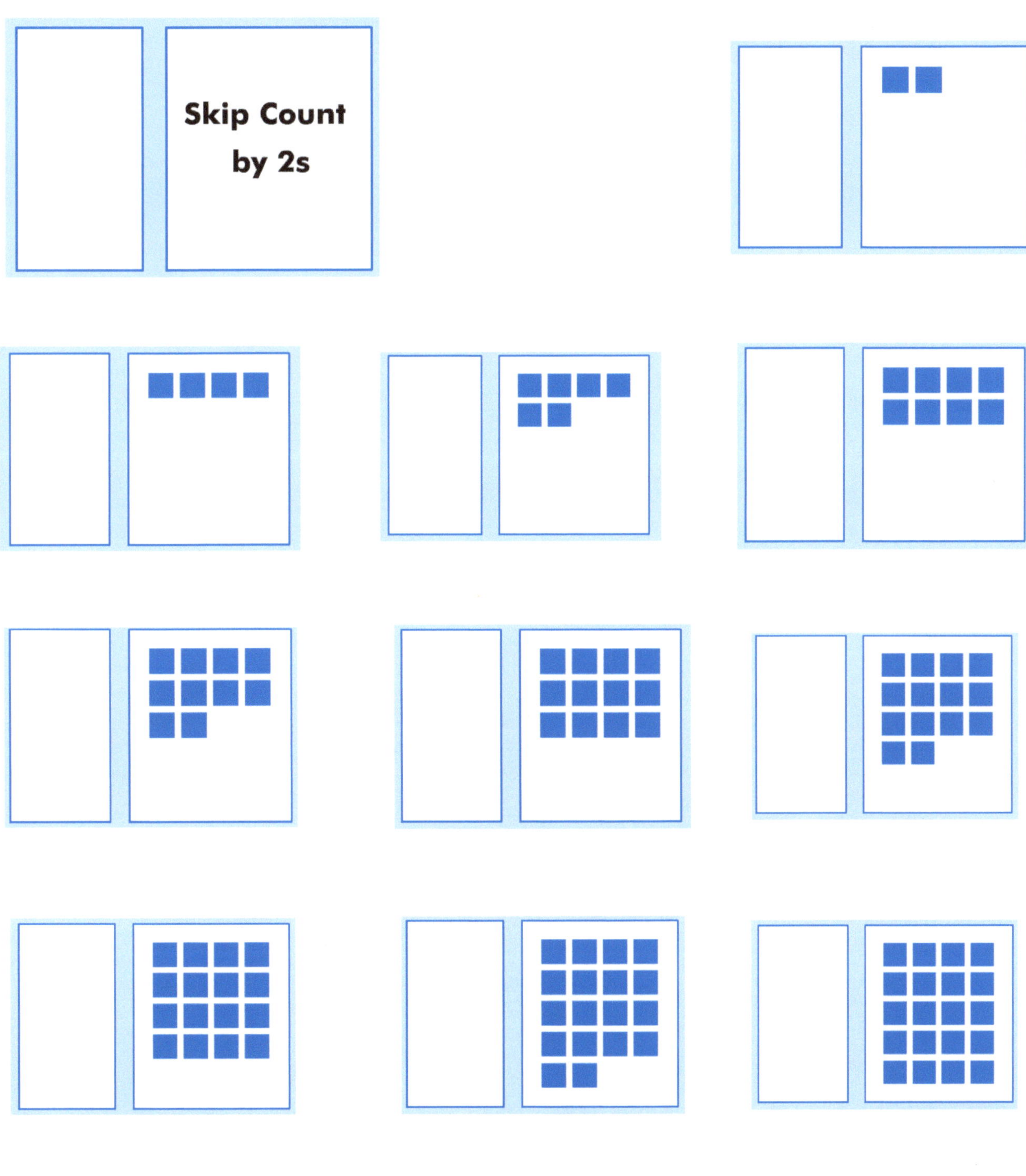

Skip Counting by Fives.

Write the number of boxes in the space provided and then count by fives.

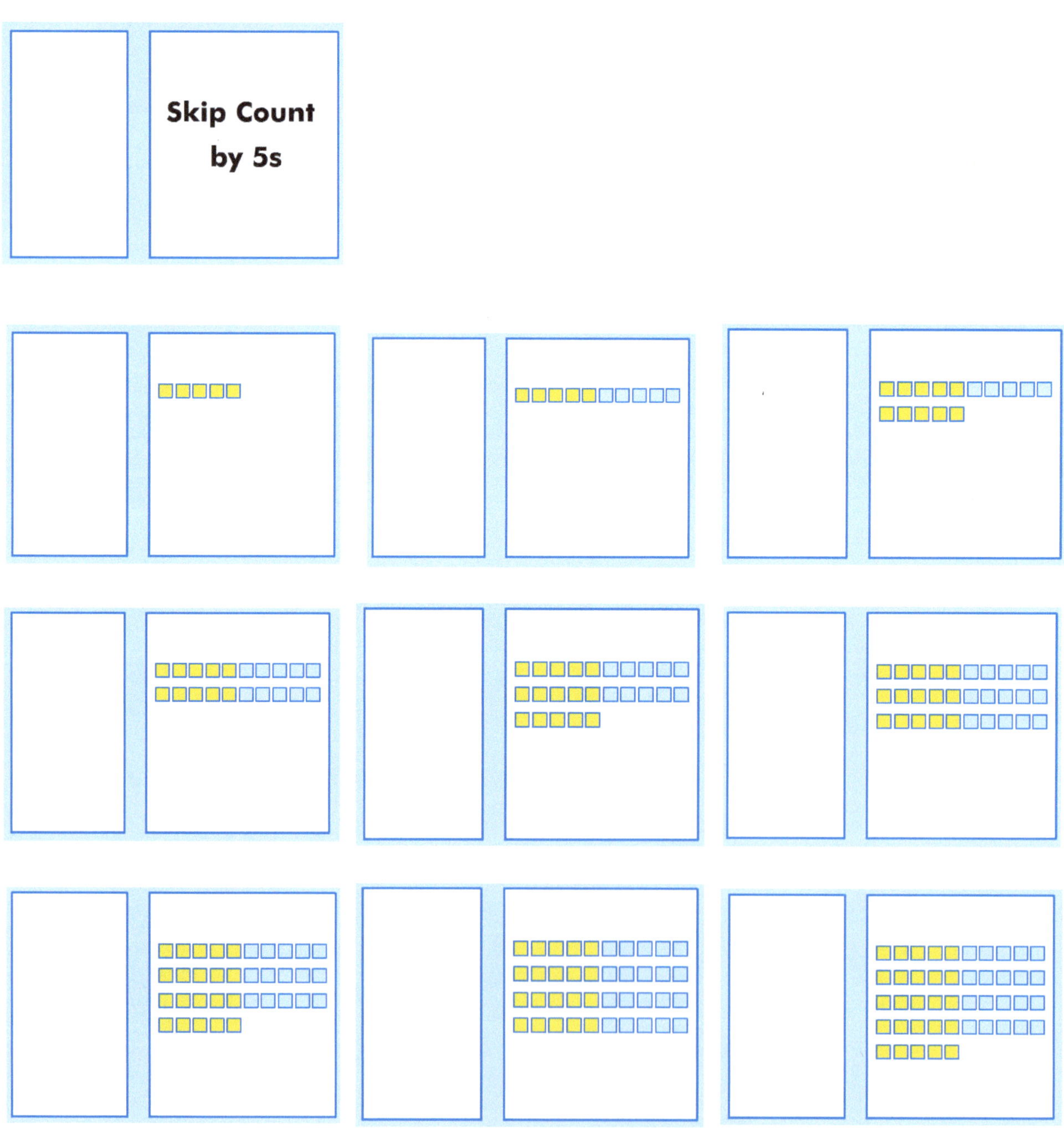

Skip Counting by Tens

Write the number of boxes in the space provided and then count by tens.

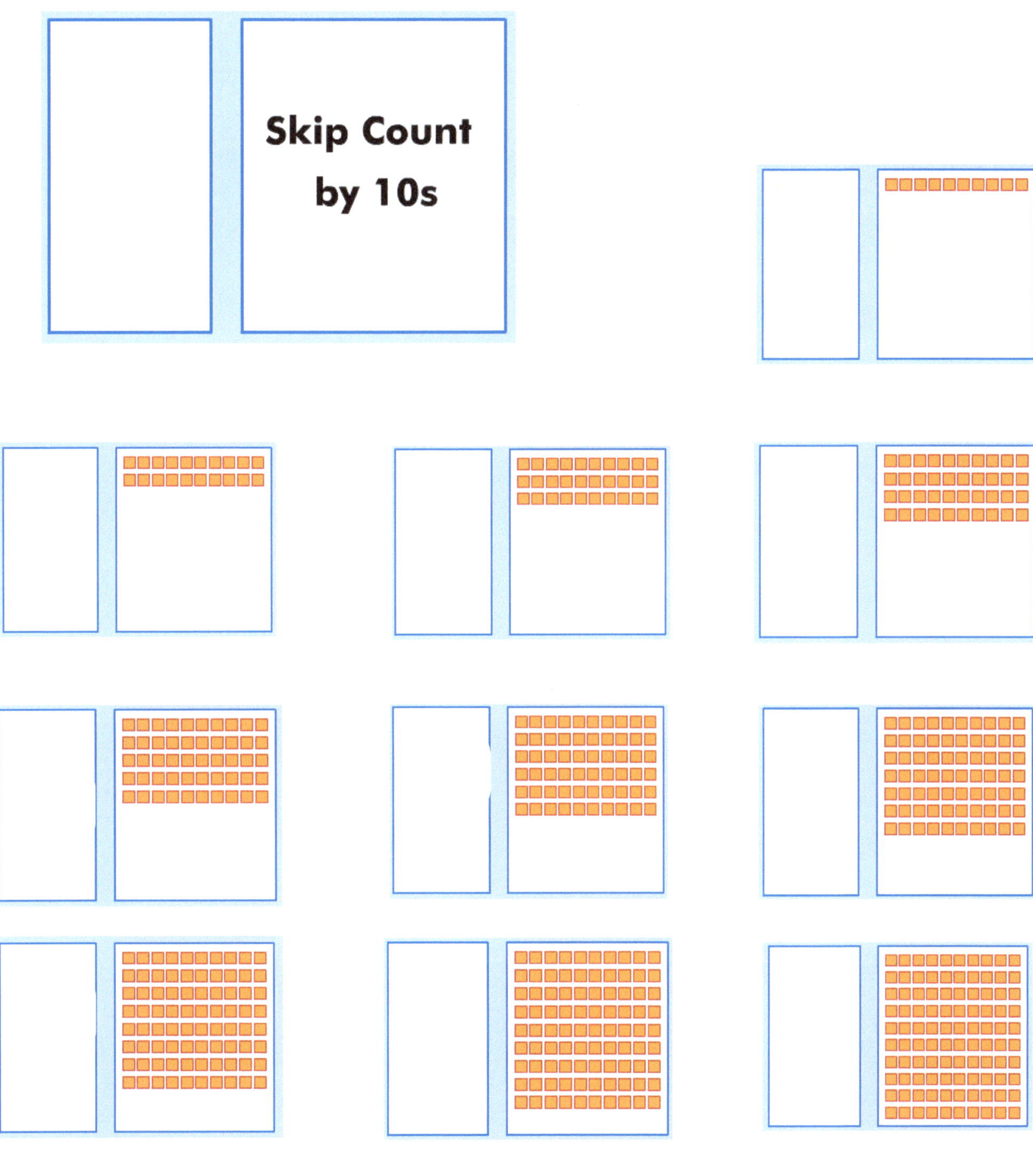

Practice by writing in the missing values.

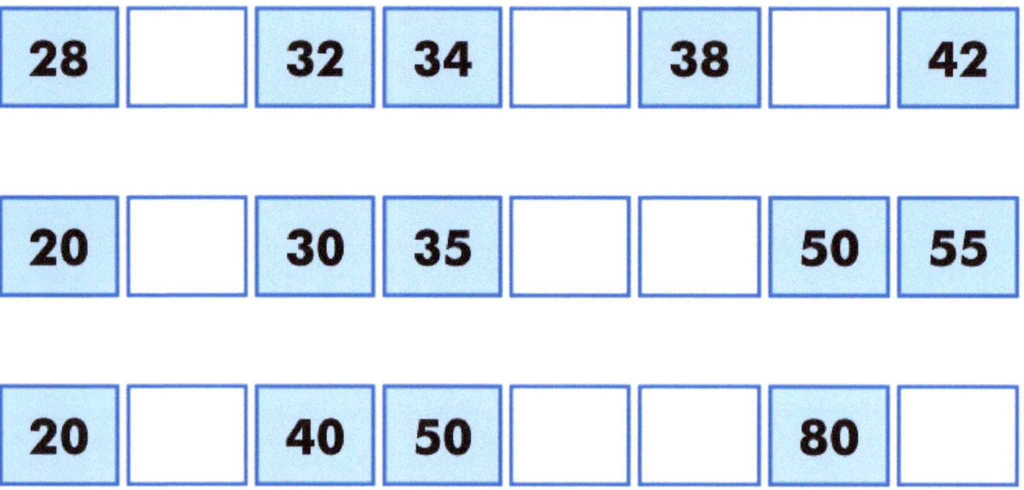

Write the red numbers in the correct positions.

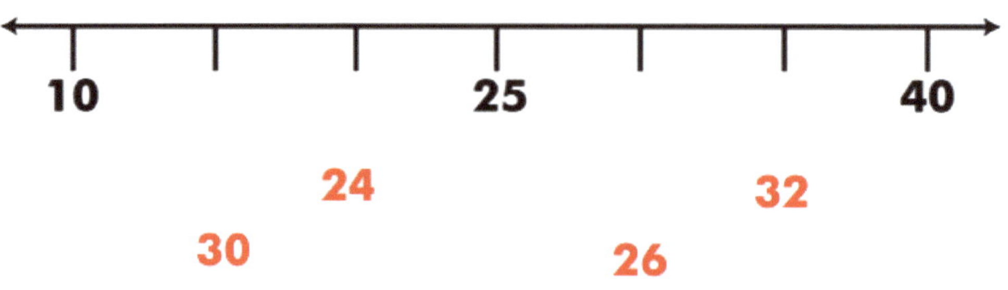

Use the 100 square to practice skip counting. Circle the numbers as you count.

Skip count by 2, 3, 5, and 10

1	2	3	4	5	6	7	8	9	10
11	12	13	14	15	16	17	18	19	20
21	22	23	24	25	26	27	28	29	30
31	32	33	34	35	36	37	38	39	40
41	42	43	44	45	46	47	48	49	50
51	52	53	54	55	56	57	58	59	60
61	62	63	64	65	66	67	68	69	70
71	72	73	74	75	76	77	78	79	80
81	82	83	84	85	86	87	88	89	90
91	92	93	94	95	96	97	98	99	100

1	2	3	4	5	6	7	8	9	10
11	12	13	14	15	16	17	18	19	20
21	22	23	24	25	26	27	28	29	30
31	32	33	34	35	36	37	38	39	40
41	42	43	44	45	46	47	48	49	50
51	52	53	54	55	56	57	58	59	60
61	62	63	64	65	66	67	68	69	70
71	72	73	74	75	76	77	78	79	80
81	82	83	84	85	86	87	88	89	90
91	92	93	94	95	96	97	98	99	100

1	2	3	4	5	6	7	8	9	10
11	12	13	14	15	16	17	18	19	20
21	22	23	24	25	26	27	28	29	30
31	32	33	34	35	36	37	38	39	40
41	42	43	44	45	46	47	48	49	50
51	52	53	54	55	56	57	58	59	60
61	62	63	64	65	66	67	68	69	70
71	72	73	74	75	76	77	78	79	80
81	82	83	84	85	86	87	88	89	90
91	92	93	94	95	96	97	98	99	100

1	2	3	4	5	6	7	8	9	10
11	12	13	14	15	16	17	18	19	20
21	22	23	24	25	26	27	28	29	30
31	32	33	34	35	36	37	38	39	40
41	42	43	44	45	46	47	48	49	50
51	52	53	54	55	56	57	58	59	60
61	62	63	64	65	66	67	68	69	70
71	72	73	74	75	76	77	78	79	80
81	82	83	84	85	86	87	88	89	90
91	92	93	94	95	96	97	98	99	100

Name:____

Skip Counting Quiz

Circle or fill in the correct answer.

1   **True or false, the missing number in this skip counting sequence is 30?**   **5, 10, 15, 20, 25, ___, 35**

2   **Which skip counting sequence is not correct?**

A   **12, 14, 16, 18, 20**

B   **35, 40, 50, 60, 70**

C   **2, 4, 6, 8, 10, 12**

D   **20, 30, 40, 50, 60**

3   **2 x ___ = 16**

4   **5 x ___ = 25**

Place Value

Key Vocabulary

Digit

Ones

Tens

Hundreds

What value is represented by each of these blocks? Write the answer in the box.

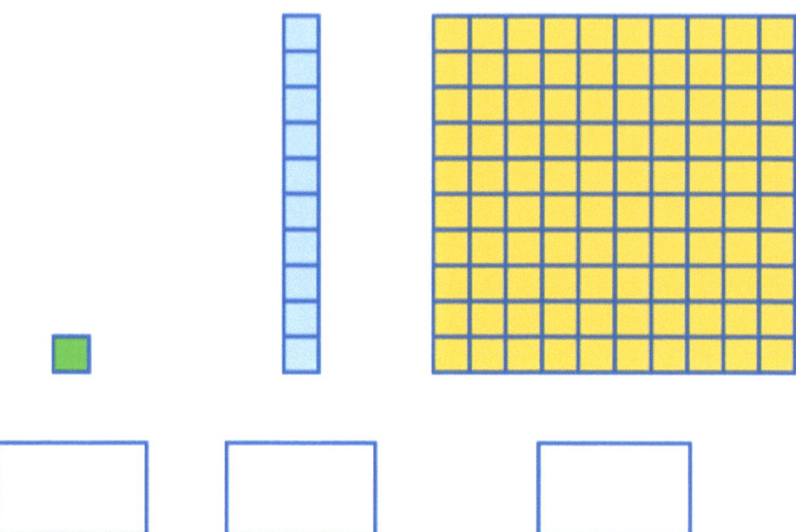

What number do the blocks represent? Hint; count the number of rows of ten blocks and then the number of single blocks (green).

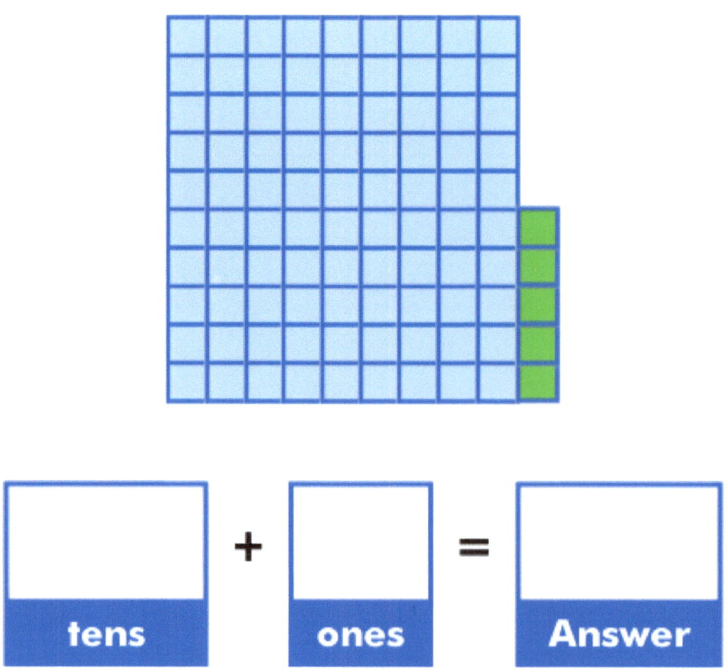

What number do these blocks represent?

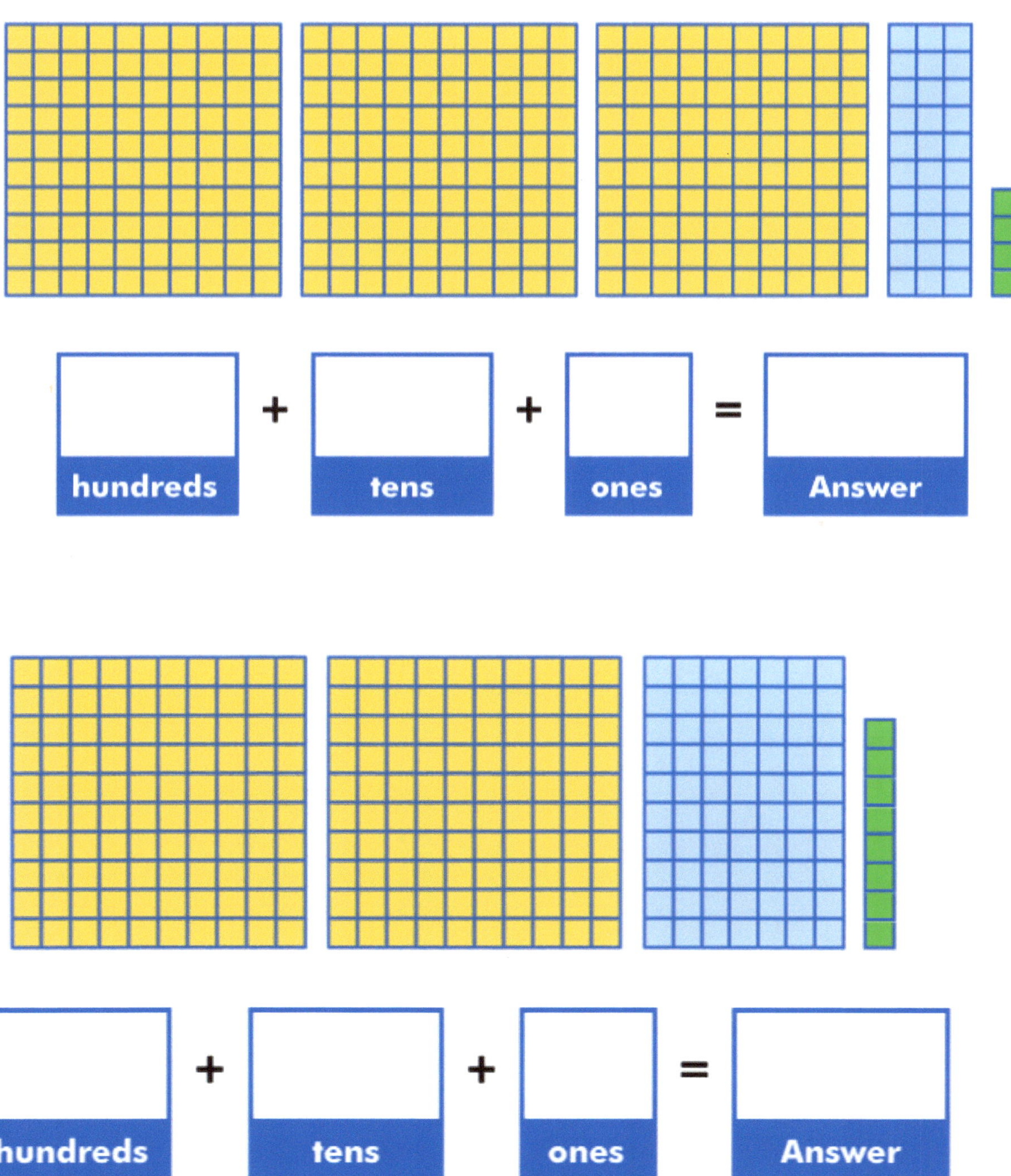

How many of each of the 100 blocks, ten blocks and ones blocks do you need to make these numbers? Write the answer in the space provided.

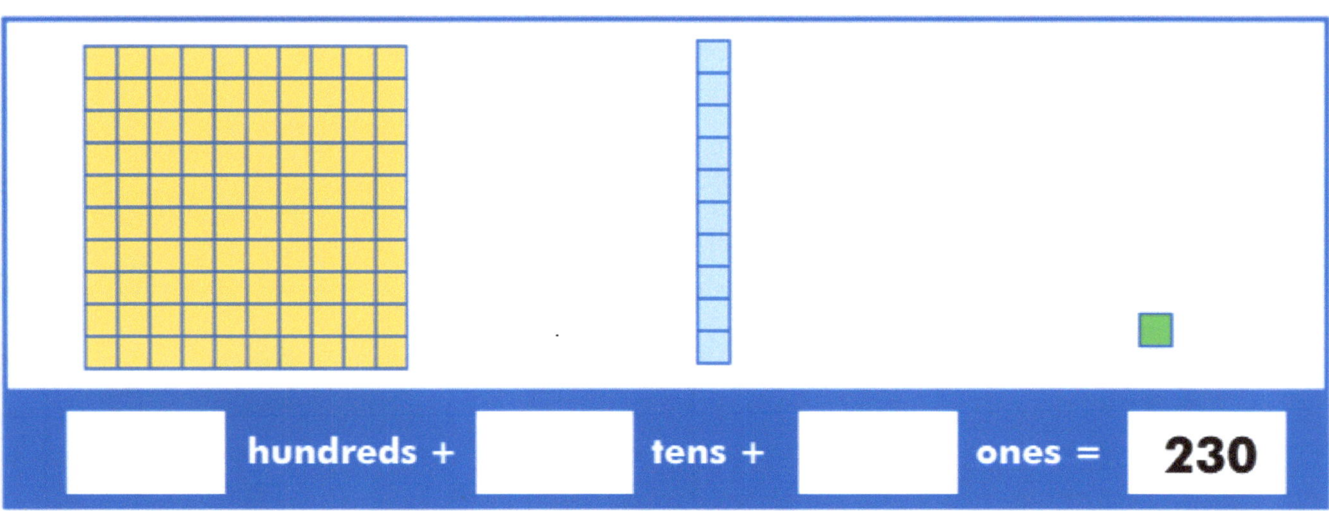

hundreds + [ ] tens + [ ] ones = **230**

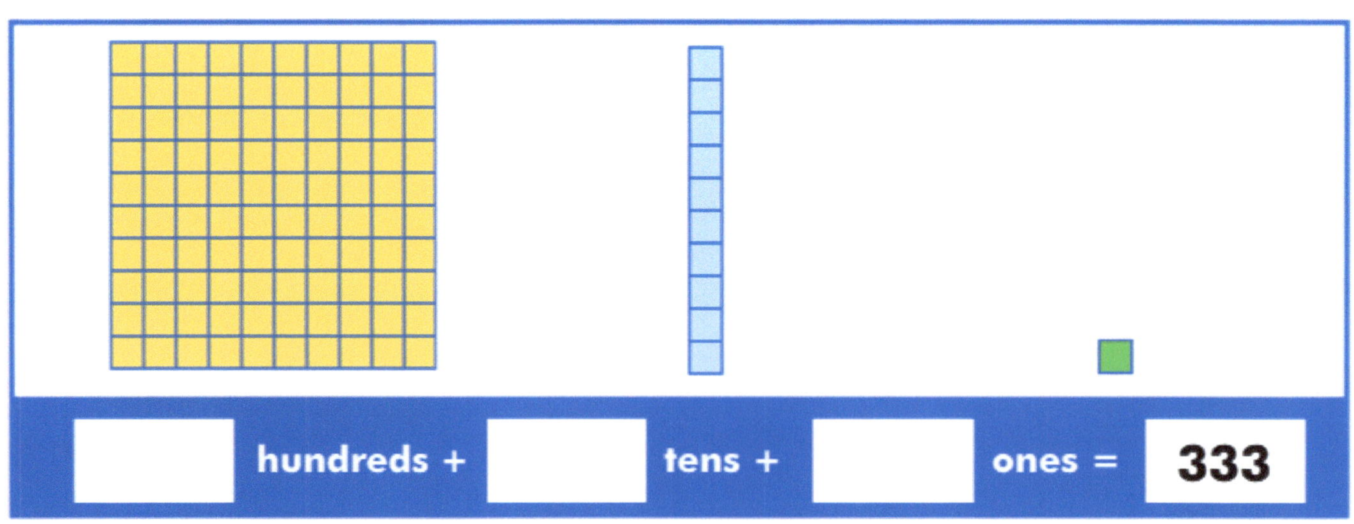

hundreds + [ ] tens + [ ] ones = **333**

What is the value of the red 5? The first one is done for you.

3**5**7   27**5**   **5**70

50

Name_______________________________________

Place Value Quiz

Circle or fill in the correct answer.

1 **8 6 5** = 80 + 60 + 5

2 **What number do the blocks in Figure 1 represent?**

(A) 66

(B) 68

(C) 58

(D) 108

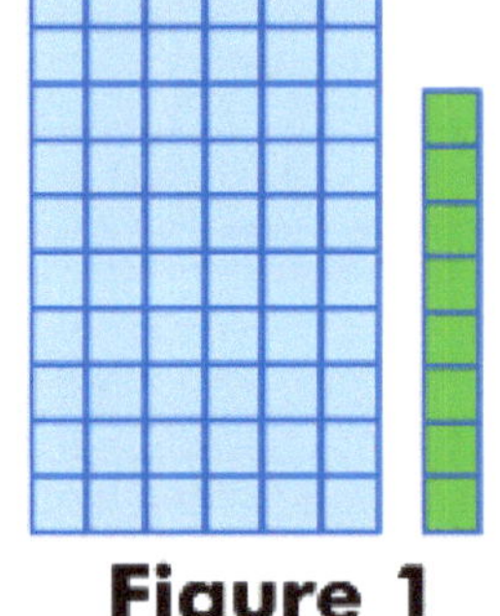

Figure 1

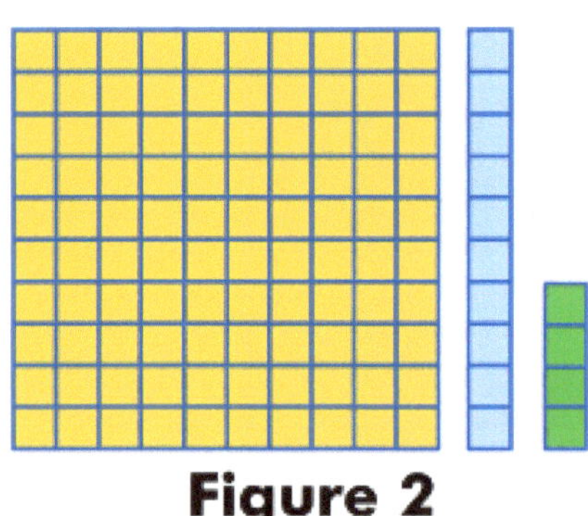

Figure 2

3 **What number do the blocks in Figure 2 represent?**

4 714 = 700 + ____ + 4

Odd & Even Numbers

Key Vocabulary

Pairs

Odd numbers

Even numbers

Sort the odd and even cards. Draw the cards in the empty boxes.

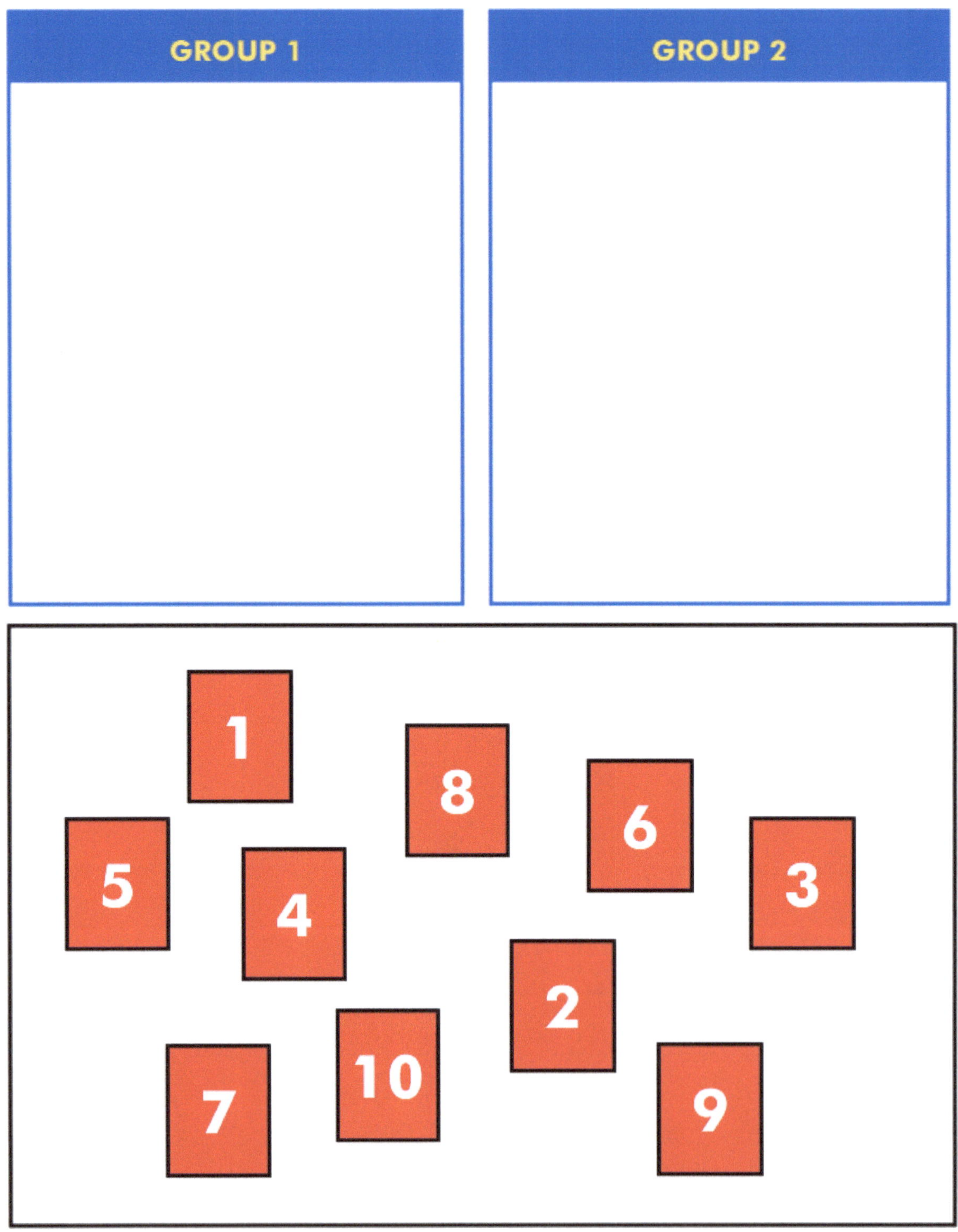

Partner up the students by connecting them with a line.
Any student who doesn't have a partner, connect them with the green box.

Draw a line between the number and the corresponding number icon.

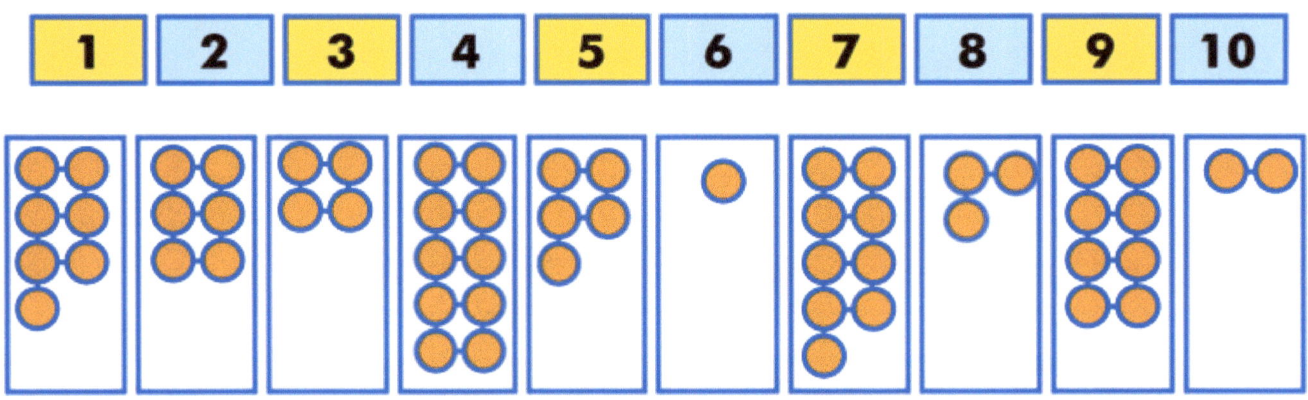

Why do you think these numbers are called "odd" numbers?

1	3	5	7	9

____

____

Why do you think these numbers are called "even" numbers?

2	4	6	8	10

____

____

Find the missing numbers and write it in the space provided.

Odd or even? Write the number in the proper box.

Odd Numbers   1, 3, 5, 7,  11, 13,  17, 19, 

Even Numbers   2, 4, 6, 8,  12, 14,  18, 20, 

ODD NUMBERS	EVEN NUMBERS

52   29   32   100   41   71

28   57   46   13   94   63   95   44

What do you notice? Fill in the blue and gold boxes.

Odd numbers always end in 

Even numbers always end in 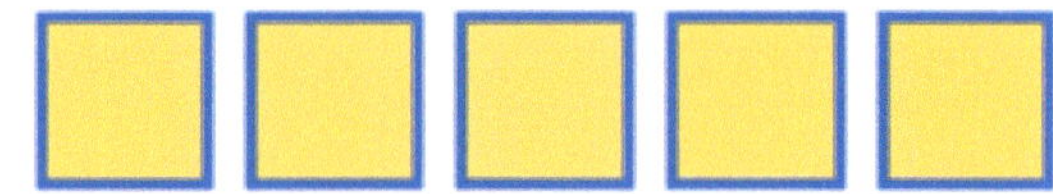

Name____

Odd & Even Quiz

Circle the correct answer.

1 **True or false? There is an even number of boys and girls.** 

2 **Even numbers always end with a digit of:**

A 2, 4, 6, or 8

B 0, 2, 4, 6, or 8

C 1, 3, 5, 7, or 9

D 3, 5, 7, or 9

3 **Which number is an odd number? 154 889 304 990**

4 **Which number is an even number? 71 89 65 90 111**

Ordinal Numbers

Key Vocabulary

first

second

third

fourth

fifth

sixth

seventh

eighth

ninth

tenth

What is the current position of the cars on this road? If you have crayons draw the color in the correct box, if not draw a line or write the name of the color.

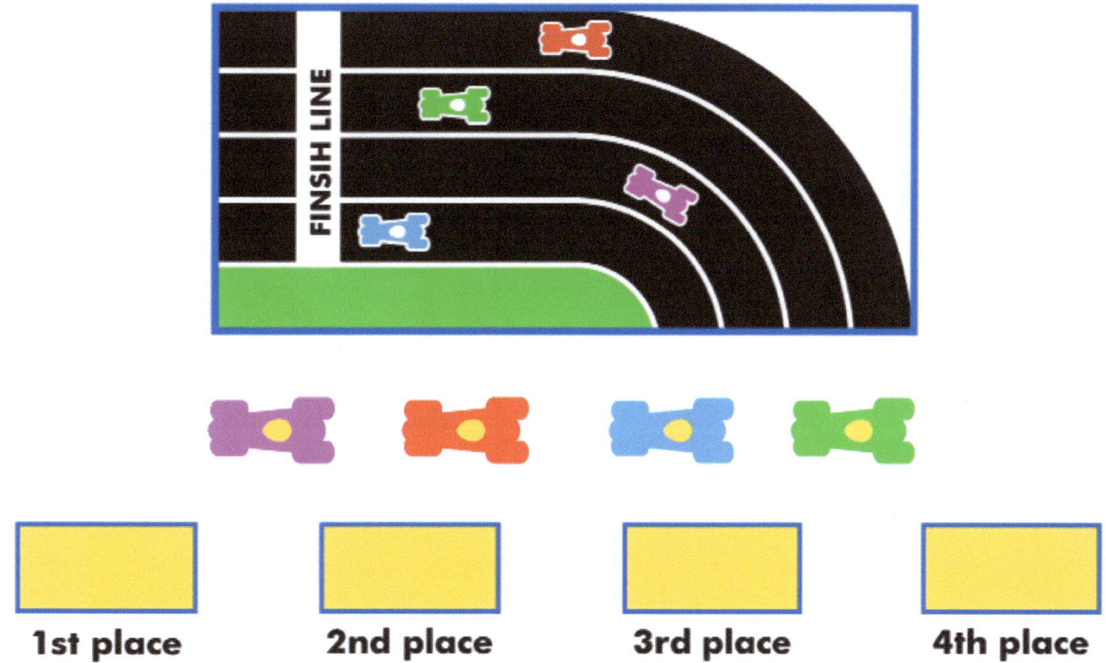

Complete this chart with the suggestions provided to see the first 10 ordinal numbers and their abbreviations. Study the chart.

Ordinal Number	Abbreviation
First	1st
Second	2nd
	3rd
Fourth	4th
	5th
	6th
	7th
Eighth	8th
	9th
Tenth	10th

Seventh

Ninth

Sixth

Third

Fifth

Draw a line from the ordinal number on the bottom to the correct date. The first one is completed for you.

By the way, do you know what is special about the date in yellow? _______________

February 2008

S	M	T	W	T	F	S
					1	2
3	4	5	6	7	8	9 Dad's 50th
10	11 Dentists	12	13	14	15	16
17	18	19	20	21	22 Business trip to Madison, WI	23
24	25 Golf	26	27	28	29	

eighth	sixth	fifth	tenth	seventh

Ordinal number word scramble.
Unscramble the ordinal numbers and then write in the abbreviation. The first one is done for you as an example.

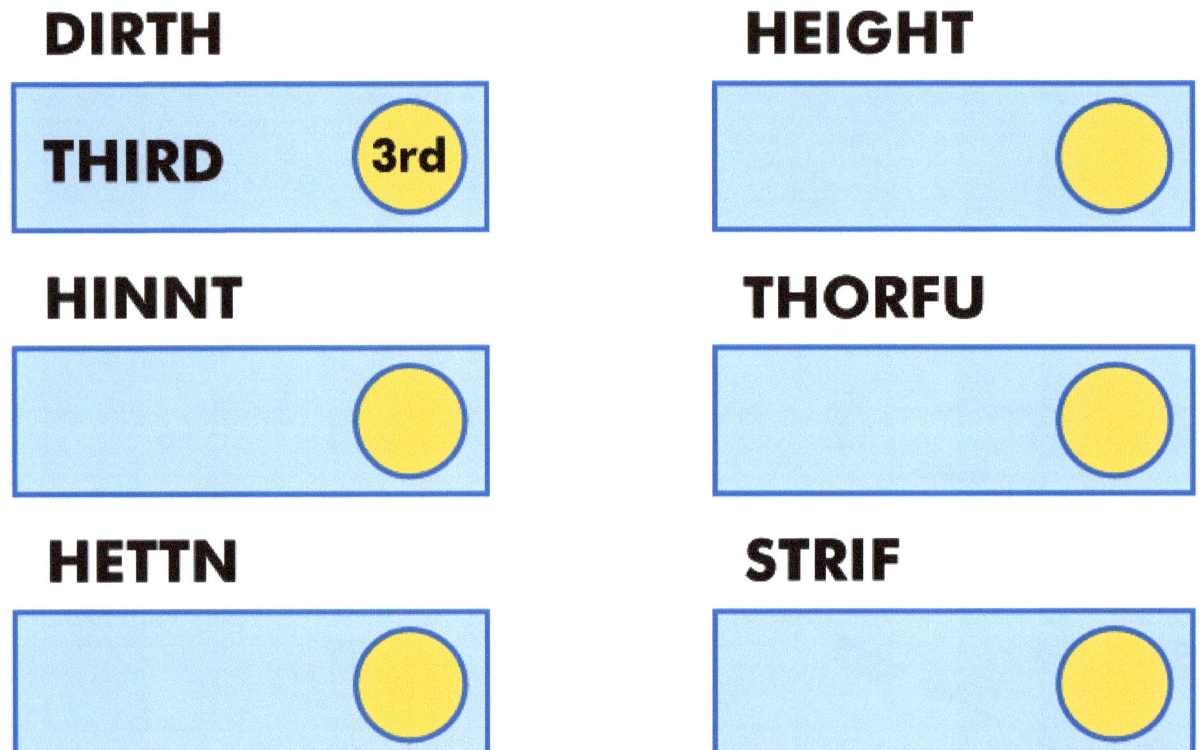

Order the students by writing in their place.
Mia came in first in the math test. What place did the other students come in?

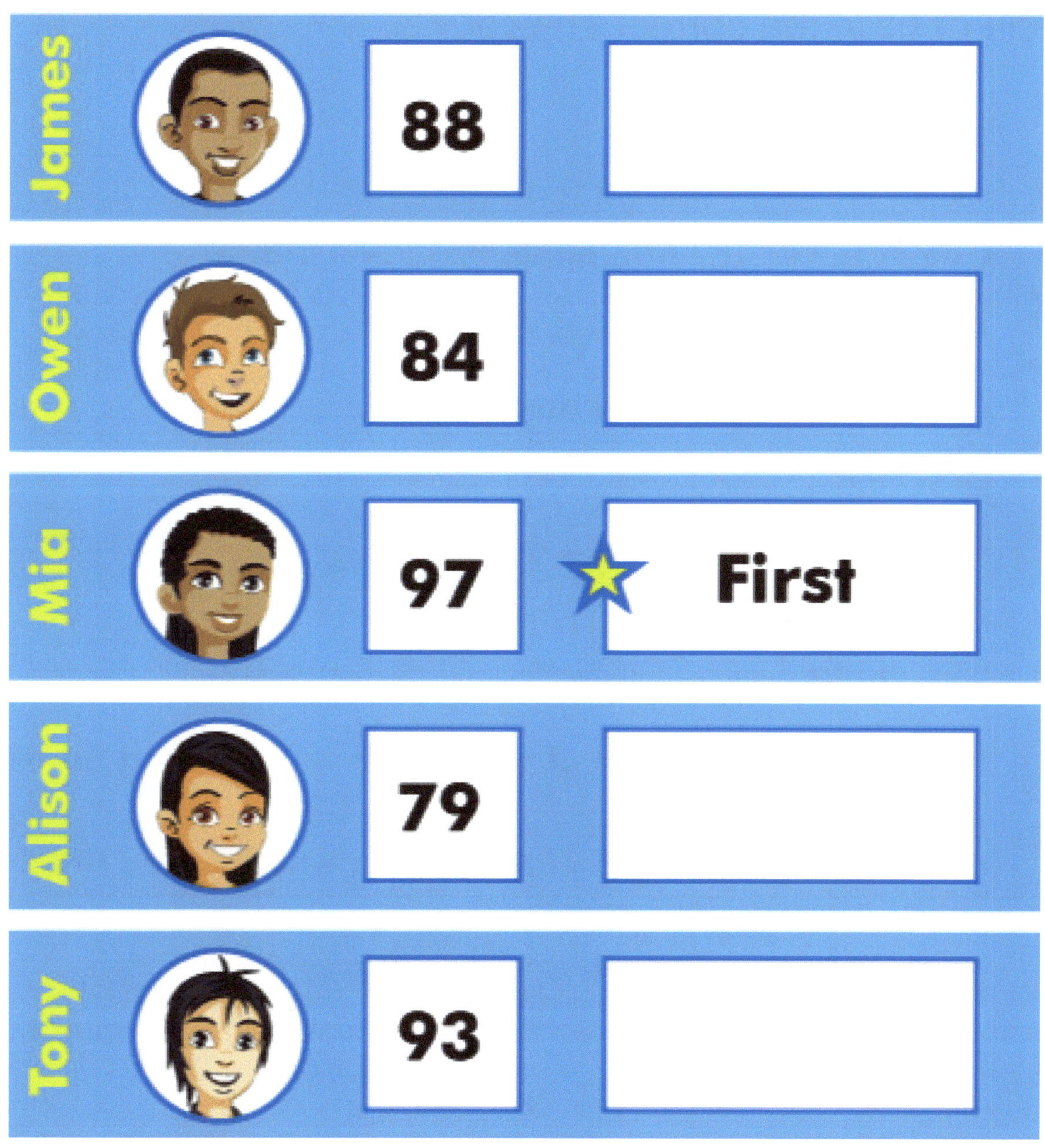

Name_______________________________________

Ordinal Number Quiz

Circle or fill in the correct answer.

1 **True or false? The fifth apple has been bitten.**

2 **Which apple has been bitten?**

   A First

   B Eighth

   C Tenth

   D Ninth

3 **What was the test score of the person who finished third?**

4 **What was the test score of the person who finished fifth?**

Test Scores	47	33	49	27	48	44

Newburyport, MA 01950

1-800-596-3175

OnBoard Academics employs teachers to make lessons for teachers! We create and publish a wide range of aligned lessons in math, science and ELA for use on most EdTech devices including whiteboard, tablets, computers and pdfs for printing.

All of our lessons are aligned to the common core, the Next Generation Science Standards and all state standards.

If you like our products please visit our website for information on individual lessons, teachers licenses, building licenses, district licenses and subscriptions.

Thank you for using OnBoard Academic products.

www.ingramcontent.com/pod-product-compliance
Lightning Source LLC
Chambersburg PA
CBHW042136030726
47599CB00002B/495